THE METAPHYSICS OF AUDIO PRODUCTION

A *Beginner's Guide to Audio Engineering Foundations*

JARON "THE SECRET" ADKISON

Foreword by
ERYKAH BADU

Melodizm Productions, LLC.

Copyright © 2023 by JaRon Adkison

All rights reserved.

No part of this book may be reproduced in any form or by any electronic or mechanical means, including information storage and retrieval systems, without written permission from the author, except for the use of brief quotations in a book review.

Paperback ISBN: 979-8-9882892-8-9

iBook ISBN: 979-8-9882892-9-6

Cover Illustration by Darien Clark.

Interior Illustration by Dedrick Strong.

Interior Illustration by JaRon Adkison.

Thank you for purchasing an authorized copy of this work and supporting all the energy invested to its fruition. Find something that evolves you within this journey and hold onto it tightly. May all that is for you be yours.

Printed by Lulu.com in the United States of America.

Melodizm Productions, LLC.

Dallas, TX

THANK YOU & DEDICATION

To The Love of My Lifetimes, Erica "Erykah Badu" Wright

Thank you for mentoring me, encouraging me, supporting me, inspiring me, teaching me, helping me to gather all this information, helping me to organize it, bouncing ideas with me back and forth, making all those trips to the printer, sorting through all those drafts, each glass of juice and every cup of tea. There is no way I would have finished this without your gentle, consistent nudges and firm support. You are a super genius and it is my honor to share this with you. I love you.

To My Village, My Mothers and Fathers, Siblings and Cousins, Aunts and Uncles, All My Ancestors, Willie Andrews, Joe Beats, Floyd "Butch" Bonner, James Dumain, Eddie Simmons, Wayne Stalling, Jonathan "Jay T" Thomas, Malcolm "Mac The Man" Wilson

I've learned so much from you in so many different ways and I look forward to so much more fun with y'all.

*May our Family that has transitioned to the next plane
Rest In Peace and Love.*

This book is dedicated to Clara Adkison, James Adkison, Brandon B. Bush, Christopher "CTK" Kelly, Marquinn Polk, Gregory "Sherb" Sherbit, Ralph "Phantom" Stacy, Larry D. Woodard and all those who inspire us to be much greater from beyond the veil of mortality.

Thank you. We love you.

TABLE OF CONTENTS

Foreword — vii
Preface — ix

1. Measuring Frequency — 1
2. Frequency and The Human Body — 5
3. Introduction to Signal Processors and Effects — 8
4. Equalizers — 10
5. Audio Dynamics — 19
6. Types of Compressors — 28
7. Working With Your Sound Field — 34
8. Limiting — 42
9. Dithering — 47
10. Signal Path (Flow) — 49
11. Gain Staging — 53
12. De-Essers — 55
13. Compression: Downward vs. Upward — 57
14. Gates — 59
15. Effects Processors — 61
16. Microphone Types — 63
17. Creating Your Chain — 69
18. Some Tips — 72
19. Outro — 77

Space For Notes — 79
About The Author — 81

FOREWORD

Defining Metaphysics:
Metaphysics is the branch of philosophy that studies the first principles of being, identity and change, space and time, causality, necessity and possibility. It includes questions about the nature of consciousness and the relationship between mind and matter.

> **Synonyms**: epistemology, cosmology, mysticism, religion, philosophy, ontology, metaphysic, phenomenology, aristotelian, theory-of-knowledge and metaphysical
>
> **Antonyms**: physical, physiological, palpable, external, substantial, practical, corporal, material, objective

Let's explore the unique, molecular relationship that happens between the engineer and his music.
Where do they meet?

Simply put, where the technique and instinct meet is where you will find the metaphysics of audio production. This is the relationship between mind and matter. If you are reading this, obviously you ARE

FOREWORD

made of the right elements required to be an audio engineer. Something brought you here.

There are 5 simple properties you apparently already possess that are required to be a good sound engineer. They are as follows:

- A rare love for music and sound . ✓ Check.
- A knack for organization . ✓ Check.
- A free swinging metronome for adventure, hard work and learning. ✓ Check.
- A general understanding for NOTES AND TIME. ✓ Check
- And the will to see another's vision through. ✓ Check.

There are two worlds that co-exist in the musical universe. The fantastic physical world of NOTES and the hidden ethereal world of SPACE or TIME. Both these worlds branch out into mini universes of endless possibilities. For the purpose of this book, NOTES can be defined as the sound that the "physical" ear hears and SPACE can be defined as what the "etheric" ear hears/feels. This can also be described as the breath or silence underneath it all. The Merging of both worlds greatly depend on the CHOICES made by you, the engineer.

This is the relationship between mind and matter. There are thousands of elements in our physical world, some of them discovered while others float about waiting to be recognized by you. Every time we hear an idea in our heads, a network of complex mathematical formulas are at work. Imagine that we connect to just the right one.

Everything is vibration. Everything is always moving. Everything is connected. All things work together in some way to help create the next thing. The atoms in the body rotate at the same rate, in the same direction as the earth. We are connected to one another in a similar manner.

Erykah Badu

PREFACE

The purpose of this book is to give you some information on energy and audio equipment. Once you receive this information, what you do with it is completely your choice. Let's get rolling.

I

MEASURING FREQUENCY

"In the beginning there was nothing, which exploded." - **Terry Pratchett, Lords and Ladies**

To begin, each sound has its own frequency response, which is the overall range of frequency that can be heard in the signal (and the amount of each frequency relative to one another). Basically, each sound has its own amount of LF, LMF, MF, HMF and HF, its own frequency response. When we record and play back our instrument, we can look at our frequency analyzer to see the 'response' of the instrument. This goes for single instruments, sections of instruments, entire mixes and even the sounds you hear around you in the world. Everything has a frequency response.

> **NOTE:** *These acronyms are commonly used in the studio and stand for Low Frequency, Low-Mid Frequency, Mid Frequency, High-Mid Frequency and High Frequency.*

THE METAPHYSICS OF AUDIO PRODUCTION

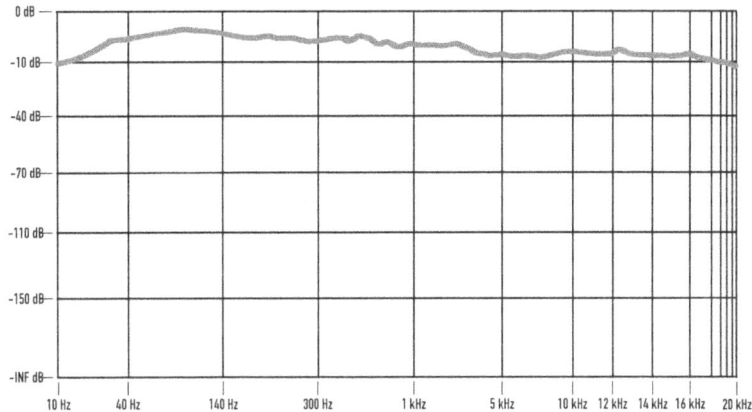

Frequency is measured in Hertz (Hz) on a frequency meter (known as an analyzer). The x-axis, read from left to right measures the frequency, from lowest to highest. The y-axis, read from the bottom to the top, measures the volume of each frequency in decibels (dB). According to the science, the human ear can only hear between 20 Hz and 20,000 Hz (20 kHz). This is the range you'll see on most analyzers.

Frequency Response

To give you an idea, 808's and bass are usually around 20-40 Hz on the low end side. They do have MFs and HFs though. Depending on the sound, the amount of mid and high frequencies will differ.

Vocals on the other hand don't have as many lows as bass instruments. Depending on the mic, vocalist and recording area, you can catch signals as low as 65 Hz or so. On the high end side, vocals reach up to 16 kHz (16,000 Hz) and so on.

2
FREQUENCY AND THE HUMAN BODY

"I was thrown out of college for cheating on the Metaphysics Exam; I looked into the soul of the boy next to me." - **Woody Allen**

Let's talk a little about sound and the Chakras. The ones we'll work with in this book are:

Root
Your sense of safety; Tribe.
396 Hz - I AM - Red

Sacral
Your sense of Creativity and Sexuality.
417 Hz - I FEEL - Orange

Solar Plexus
Your Drive, Will; What you will and won't do.
528 Hz - I DO - Gold

Heart
Your inner most, truest feelings.
639 Hz - I LOVE - Green

Throat
Your voiced opinions; What you're willing to say.
741 Hz - I TALK - Blue

Third Eye
Your understanding of the world around you; The Layers of Existence.
852 Hz - I SEE - Indigo

Crown
Your connection to the Higher Power.
963 Hz - I UNDERSTAND - Ultraviolet

THE METAPHYSICS OF AUDIO PRODUCTION

The energy centers in our bodies are tuned to specific frequencies that can be stimulated by a number of instruments. A sound therapist understands the way this relationship works, mixing specific instruments and playing techniques with a self-reflective, meditative practice to help enhance the innate healing capabilities of the human body.

With that said, have you ever heard an 808 drop and just started dancing? Have you ever witnessed a cypher in a lunchroom where lyricist after lyricist rap over someone beating on the table with just their hand and a pen? Have you ever got a free back massage listening to music in a car with additional subwoofers in the trunk? If we consider this understanding with respect to the material we're working on, we can enhance the sonic arrangement of the mix and the overall experience of the piece's storyline.

3

INTRODUCTION TO SIGNAL PROCESSORS AND EFFECTS

"Be strong - I said to my Wi-Fi signal." - **Erykah Badu**

THE METAPHYSICS OF AUDIO PRODUCTION

In the studio, an audio engineer has access to many different tools they can use to accomplish their goal sonically. In the next few chapters we'll discuss some of the most commonly used among them and go over their basic capabilities. Hardware examples are provided for each unit and there are also softwares modeled after each of these processors, available for purchase online from various companies. These software emulations are known as 'plug-ins'.

NOTE: *Plug-ins are digital emulations of analog devices, designed to be used in your DAW (Digital Audio Workstation). DAWs are programs like Apple's Logic Pro X, Avid's Pro Tools and Ableton's Live. Emulations are really good imitations, that can be almost effective as the original. There are a variety of plug-ins available, from equalizers to compressors to reverbs, etc.*

In earlier history, we used mixing consoles and tape machines to record, moving into the cassette and 8-track era, followed by a legendary device known as the ADAT. The ADAT gave us the capability to record straight to a hard drive. It was one of the earliest versions of this technology. Fast forward to today and we have almost completely digital studios, made up of mostly computer software. Everything's happening in the DAW now, 'in-the-box' (ITB). Everything's a plug-in now, so let's talk some history on the hardware that changed the game and inspired so many new ways to do what we do. Coming up first, the 'Equalizer'.

❦ 4 ❧
EQUALIZERS

"The revolution will not be televised." - **Gil Scott Heron**

An equalizer (EQ), is a device designed to boost or cut specific frequencies in an audio signal (sound). We can use EQs to remove muddiness, add sheen to achieve clarity and/or drastically change the overall tonality of a sound source.

Terms like 'muddiness', 'sheen' and 'clarity' are all related to the frequencies being cut or boosted. We can cut LMFs to remove muddiness, add HFs to bring out sheen (glossiness) or add even higher HFs to enhance clarity (clearness of the high end). This all depends on mixing style, though there are some technical rules-of-thumb that'll help you a lot.

There are two main types of equalizers and many variations of these two. Before we get into that let's talk about EQ bands.

EQ Bands and their Shapes

An EQ band, sometimes called 'frequency band' is a range of frequencies. For example, 20 Hz - 120 Hz is a LF (Low Frequency) band. Each band has a bandwidth control, sometimes called the 'Q'☐, short for 'quality factor'. Bandwidth does like the name suggests and controls the width of that particular band, which changes how wide the range of affected frequency is. The middle point of each band is called the 'center'. The higher your bandwidth, the tighter or more narrow your band will be. The lower your bandwidth, the wider or more broad your band will be. On some equalizers, you have a shape or curve control for each band. In this book, we'll cover five of them.

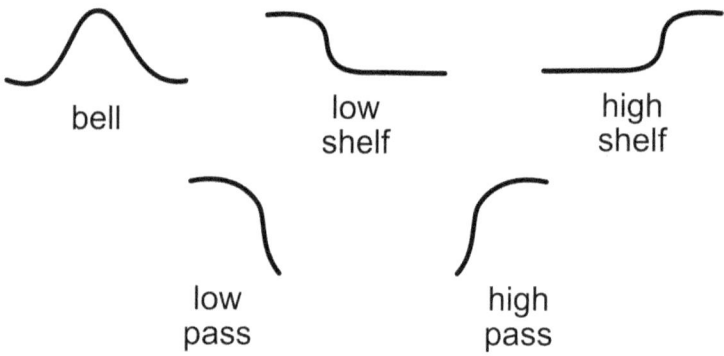

Bell - the center of the band is boosted along with the neighboring frequencies

Low Shelf - every frequency below the center of the band is boosted

High Shelf - every frequency above the center of the band is boosted

Low Pass (a.k.a. *Hi Cut*) - removes all frequency above the center of the band a little at a time as the cut extends. This depends on the sharpness of the curve.

High Pass (a.k.a. *Lo Cut*) - removes all frequency below the center of the band a little at a time as the cut extends. This depends on the sharpness of the curve.

The Graphic EQ

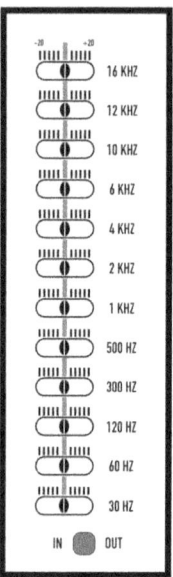

Graphic EQs have set center frequencies and 'Proportional Q', meaning that each bandwidth starts low (wide) and as you boost or cut more gain, the bandwidth gets higher (narrow).

In this example, there's a band for 30 Hz, 60 Hz, 120 Hz, 300 Hz, 500 Hz, 1 kHz, 2 kHz, 4 kHz, 6 kHz, 10 kHz, 12 kHz and 16 kHz. This type of EQ was popularized with the API 560 by API in 1967.

Parametric EQ

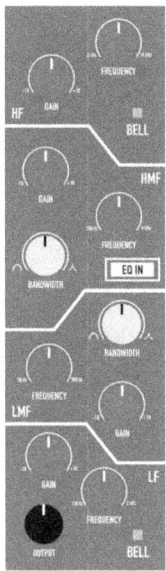

Parametric EQs generally have three or more frequency bands and often use bell curves. There are usually four bands you can boost or cut: LF, LMF, HMF and HF.

Sometimes these EQs have the option to switch the LF and HF bands from bell curves to shelf curves. There's also an Output Level control. This type of EQ was popularized with the SSL 4000 EQ in 1979.

EQ in Action

Let's talk about how to use EQ a little bit. For this example, we're going to use a parametric EQ on a lead vocal.

Here's an example of our curve.

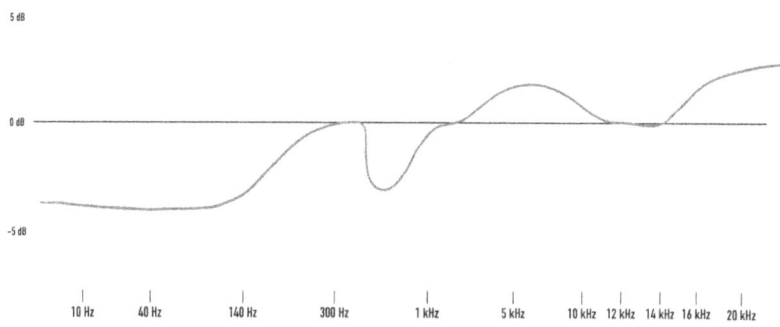

We have two Shelves for our LF and HF bands, then we have two Bells for our LMF and HMF bands. That's four bands total to work with on this processor.

We subtracted around -4 dB of 60 Hz (and below) with our LF shelf. This lessened some of our room noise in this particular recording. We subtracted about -3 dB with a pretty sharp bell around 600 Hz to clear up some mud.

We boosted about +3 dB around 16 kHz (and above) with our HF shelf. This gave our vocal more air. After that we boosted about +1.5 dB around 5 kHz with a medium-wide bell. This made our vocal more present in the mix.

> We can use equalizers different ways to clean up or change the harmonic blend of our signals.

NOTE: *Harmonics are the balance of frequencies in a specific sound source.*

They're a lot like colors. You can mix them and make new shades or even other colors. Using harmonics, we can find many new ways to express our ideas sonically.

Using Filters

Sometimes EQs come with hi and lo pass filters built in. This will help a lot with the placement of sounds in your mix. Say we're still working on the same lead vocal and we add a hi pass filter set around 100 Hz. This will help tie up a good amount of our low end vocal issues.

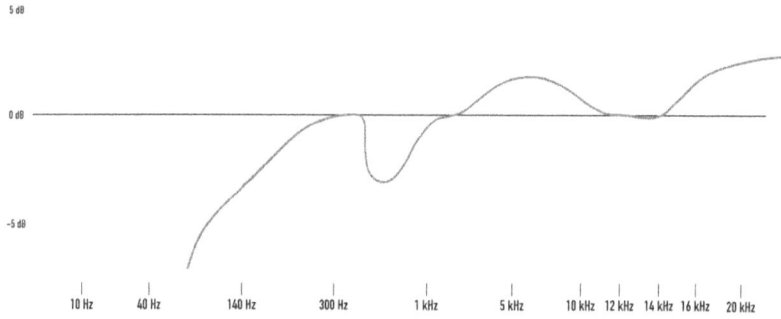

5
AUDIO DYNAMICS

"No camera can fully capture
The face I see in the mirror
There is a distortion in the energy
Like a song sung when no-one is
Listening
In a forest made of
Knives" - **@behind-the-veil-of-sanity**

First things first, let's talk about one of the most important pieces of equipment in your studio.

The Preamplifier
The preamplifier, also known as 'the preamp' is a handy device. When recording using microphones there are a couple things to be aware of. Firstly, we have to consider that microphones operate at Mic Level, which is about 1 thousand times less powerful than Line Level.

A lot of the instruments and processors we use like amps, synthesizers, drum machines, compressors and equalizers all work at Line Level. This means without a preamp our instruments could be more powerful in the mix than our vocals, which are the centers of our records a lot of the time. This does depend on style but preamps are helpful in general.

Microphone preamplifiers do as the name suggests and amplify the signal, allowing us to use the most of our signal processors and get the best possible sound out of them.

Preamps help increase the 'body' of our vocals. Body or 'Width' and 'Depth' is what makes our recordings have more space in the mix before we touch them. You can also plug instruments like a lead or bass guitar into a preamp for improved width, depth, presence, color and texture.

Afterwards, there's 'Gain Staging' and fader volume. We'll talk details about Gain Staging in Chapter 12. Fader volume is how high or low the fader is in the mix, which increases or decreases the level of that channel/track and the audio on it.

With the help of a preamplifier, you can bring your recordings closer to a commercially competitive standard.

Let's dig a little deeper into preamps and get a little more info.

'Phantom Power' (*known as +48V or P48*)
As an easy rule of thumb, majority of the microphone types available need power. Phantom Power is the standardized DC current used to power microphones. Aside from dynamic microphones, if the mic doesn't come with an external power supply (a box you plug into a wall outlet) then it needs power from a preamp.

Proximity Effect
In a recording session with a vocalist, it's helpful to be aware of 'Proximity Effect'. It's when the P's and B's in a vocal take come across the microphone very aggressively. There may even be distortion. There are a couple other reasons why this happens, but a main one is the vocalist's proximity (distance) from the microphone. This can also be caused by boosting your vocal too much with the preamplifier.

Find the right spot for your microphone in the room. Find the right settings and then depending on the style of song, work with the vocalist or instrumentalist to find the right proximity for the right sound.

Preamp Types
There are two main types of preamps we can use in the studio.

Vacuum Tube (or *'Tube'*) - known for color and warmth (takes time to warm up, at least 20 mins)

Solid State - known for transparency and clarity

Some Common Preamplifier Parameters (Controls)

There's a wide variety of preamps, made by different brands and manufacturers with different designs and tonalities. Majority of them have similar controls. The most popular two you'll see are:

Gain (also known as *Level*) - how much the preamp is boosting the volume of our signal

Drive - how much saturation we're getting from the preamp, which can also add volume

NOTE: *We can think of saturation like excitement of color. You know how you can change the picture color on your TV? With the right setting everything stands out. It can be dramatically effective.*

Dynamic Range

In audio, dynamics are the varying levels of energy in a sound source. There's a variety of dynamic processors created to alter the dynamic range of audio signals. The 'Dynamic Range' (also known as *DR*) is the difference between the lowest and the highest levels of energy in the signal.

The dynamic range for most human beings' hearing is about 120 dB. That means after 120 decibels, you probably won't be able to stand any more volume (unless that's your thing).

Be careful with your ears. They're your mixing eyes. Human hearing naturally degrades over time so take it easy and they'll return the favor.

Let's go over an example of dynamic range when mixing.

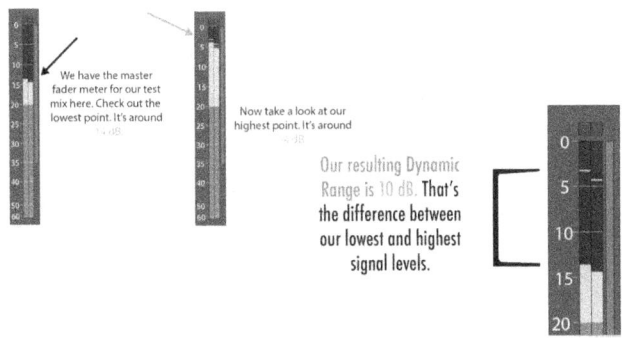

Now that we have a little info on dynamic range, let's move on to our second processor.

The Compressor

The compressor is a device designed to reduce the dynamic range of an audio signal.

Using compression to reduce the dynamic range of a signal isn't the same as lowering its volume. What happens is the compressor brings the peaks of a signal to a controllable 'threshold'. Since the highest point of the signal isn't as high anymore, the dynamic range is less than it was pre-compression.

Let's say we have a mix with 8 dB of dynamic range that's peaking at around -4 dB. We slap a compressor on, dial in our settings and lose about 3 dB of gain from the compression. At this point, our dynamic range has changed from 8 dB to 5 dB. This process of signal reduction is called 'Gain Reduction' (or 'Peak Reduction'). This is the way we monitor how much compression is happening.

This changes the volume as well as the amount of power that's pushing out the speaker. This is why 'Makeup Gain' controls are built into compressors. It helps with Gain Staging, one of the most powerful tools in your arsenal.

THE METAPHYSICS OF AUDIO PRODUCTION

Common Compressor Parameters Explained

Next, let's expand on a few of the commonly used controls on a compressor. We'll need this info later when we get into some other dynamic processors and their uses.

Ratio - the amount of compression

For example, a ratio of 4:1 (pronounced "4 to 1") means that for every 4 dB detected, only 1 dB is heard coming out of the compressor. A ratio of 6:1 means that for every 6 dB detected, only 1 dB is heard coming out of the compressor.

Attack - how long it takes for the compressor to start compressing

Release - how long it takes the compressor to stop compressing

NOTE: *Attack and Release are both measured in milliseconds (ms) to seconds (s).*

Threshold - the minimum dB level required for the compressor to activate

Signal > Detection > Compression

Let's throw out some settings and talk about what's going on here.

Example #1:
Ratio | 4:1
Attack | 30 ms
Release | 60 ms
Threshold | -9 dB

For Example 1, when the signal detected in the compressor reaches at least a level of -9 dB, it will take 30 ms for the compression to start. Our ratio is 4:1 so for every 4 dB of signal detected only 1 will be heard. The signal is compressed at this ratio for 60 ms before it is released and this process is started over. The cycle repeats until the compressor no longer detects audio.

Example #2:
Ratio | 6:1
Attack | 40 ms
Release | 100 ms
Threshold | -6 dB

For Example 2, when the signal detected in the compressor reaches at least a level of -6 dB, it will take 40 ms for the compression to start. In this case, our ratio is 6:1 so for every 6 dB of signal detected only 1 will be heard. The signal is compressed at this ratio for 100 ms before it is released and this process is started over. The cycle repeats until the compressor no longer detects audio.

As sound is being processed, compressors detect signal then attenuate it based on the design and settings. This happens over a period of time.

Now, let's get into one more important parameter that isn't always available but when it is, you'll want to use it.

Knee - controls the width of the threshold ranging from 'soft' to 'hard'.

With hard-knees, once the signal level reaches 0.1 dB above the threshold point, compression is applied. We can go for brighter, dynamically punchy sounds this way.

With a soft-knee, the threshold's range is opened up and allows signal before the threshold point to be compressed. This lessens the overall effect of the compression, making it more subtle. We can go for smoother, thicker sounds this way.

6

TYPES OF COMPRESSORS

"Moonlight floods the whole sky from horizon to horizon; How much it can fill your room depends on its windows." - **Rumi, The Essential Rumi**

Hello again,
In this chapter of The MAP we're going to cover a few popular compressor types and their styles.

The Optical Compressor

Optical compressors work using a light bulb and photocell. As voltage passes through the light bulb, the photocell detects it, triggering attenuation.

Because optical compressors are built with very few components, they tend to have a clean signal path and sound transparent. For example, let's use the LA-2A by Teletronix (debuted in 1962). It's a hybrid of sorts because it uses a tube for its gain stage, which gives it a unique character in the studio.

It has Gain and Peak Reduction controls along with a Limit/Compress switch. In Limit mode there's a higher compression ratio than in Compress mode. The attack and release parameters are determined by these three settings as well as music and level of signal. There's also a Metering control. You have the choice of monitoring your output at +4 dB or +10 dB, with a middle setting you can select to monitor gain reduction.

The FET Compressor

Field-Effect Transistor compressors, also known as FET compressors were essentially designed to emulate a tube sound, but with more reliability (due to the fragility of tubes and the maintenance they require).

Instead of a tube, it uses a Field Effect Transistor. The word 'transistor' is a portmanteau (blend of name and meaning) of the words 'transmitter' and 'resistor'. A transistor is a semiconductor, which means it conducts (transmits) electricity half the time while the other half of the time it is resisting electricity. Semiconductors are materials with electrical conductivity between a conductor like metal or a nonconductor (also known as an insulator) like glass.

Field Effect Transistors use electric fields to control the flow of current (our signal). Voltage triggers the amount of compression of our electronic signals based on the material and the compressor settings.

The 1176 FET compressor by Urei (debuted in 1967) is notable for its colorful and bright sound. It also has very fast attack and release times. The 1176 offers Attack, Release, Ratio, Input and Output Level controls.

The VCA Compressor

Voltage-Controlled Amplifiers, VCAs for short, are technically built into many electronic musical devices and computer softwares. Other compressor types even have VCAs in them.

In VCA compressors, VCAs are built into Integrated Circuit chips (also known as '*IC chips*'). When the audio signal comes in the compressor it is filtered through the IC chip where it is split down two paths. The first is the detector path, which determines the amount of compression applied to the material. The second is the output path, which is what we hear coming out of the compressor.

These compressors are known for being smooth and adding a nice, predictable 'glue'.

NOTE: *Glue is just another way to say cohesiveness or togetherness. When something adds glue to our mix, we hear all our sounds come together and blend into a mutual texture. They sound and feel like they're in the same space and time.*

VCA compressors are very popular on the master bus (your overall mix or master output). One popular model is the legendary SSL G Comp (released in the 1980s as part of the SSL G Series analog console).

The Tube Compressor

Tube compressors, while similar to FET in design, work using a vacuum tube as the detector for peak reduction. With this design, when voltage passes through the tube it is attenuated.
Tube compressors tend to have much slower attack and release times. They are usually really forgiving (meaning yielding to the material) resulting in a very musical style of compression and overall texture. It can be quite complementing to the right material. The 670 by Fairchild (developed in the early 1950's) is a renowned tube compressor. It offers Input, Threshold, Time Constant (which sets attack and release settings) and an option to switch between Stereo and Mid/Side modes, shown as 'Left/Right' and 'Lat/Vert'.

7
WORKING WITH YOUR SOUND FIELD

"Don't trust atoms, they make up everything." - **Erykah Badu**

THE METAPHYSICS OF AUDIO PRODUCTION

Have you ever been listening to a project on your monitors, dropped your pen, went down to pick it up and said, "Hey, I didn't notice that in the mix before."?

To keep it short, there's a nice amount of factors that affect what we hear in the studio. A few are your room's design, your acoustic treatment, your monitor placement, listening position, playback volume and your monitoring device whether it be your audio interface or a monitor controller and its quality.

When we play sound from a DAW, first it travels through the code of the computer program as data. (If you're on a Mac, then the second step is macOS' CORE audio driver.) The sound is then sent through the data cable connected to our 'audio interface', where it is converted into analog audio and sent through audio cables to our monitors. This trip will vary depending on the setup.

NOTE: *An audio interface is a device designed to connect all our sound sources and processors to our computer. It usually has mic preamplifiers built in it as well as A/D-D/A converters (which we'll talk about more in the next chapter).*

All those factors affect how we experience the 'Sound Field'. Imagine you're sitting at a desk. You take out a pencil and place it in the center of the desk. You crouch down until you can see the pencil at eye level. At this point let's say this desk represents the sound generated by both your speakers and the pencil represents a snare drum.

As we move the pencil closer to us, the snare gets perceivably louder in the mix, and the opposite occurs if we move it further away. If we move the pencil to the left, we begin to hear more snare in the left speaker than the right. The opposite happens if we move it more to the right side.

Pulling the 'fader' up and down is like sliding the pencil closer or further away from you. Turning the 'pan control' left or right is sliding the pencil to the left or right. There's even effects we can add to generate space, kind of like how the preamp increases the body of our vocals. It's the same type of concept, just a different application of it. We'll get into those effects in Chapter 15. Let's dive a little deeper into Faders and Panning for now.

Fader Volume

Faders control the volume level of a channel or audio track on a mixer. In today's time, most studios have some sort of DAW in action, so the faders are virtual. When working with these softwares we're basically using an emulation of an analog recording console with advanced editing and routing features. These consoles and their designs are a big part of the reason why we understand music production the way we do. A workflow was popularized, refined then evolved in many ways over time.

> **NOTE:** *When we're recording, we want to make sure to gain stage our signals. We want the best possible sound quality, a nice level and some headroom for further processing. After that, our fader comes into the game. Getting a good balance of your signals and/or audio clips with all your faders at 0 dB is a great way to start a recording or mix.*

Panning

Pan controls change the lateral (left and right) placement of a signal in our mix.

> **NOTE:** *It's helpful to spread things out in your mix, giving each piece of the music space and room to be expressed without taking away from another piece.*

Stereo vs. Mid/Side

Stereo is when all the signal is separated between the Left and Right channels. Mid/Side (also shown as 'M/S') is when all the signal is separated between the Middle and Side channels. I have a couple graphics to visually explain what I mean.

Stereo Separation

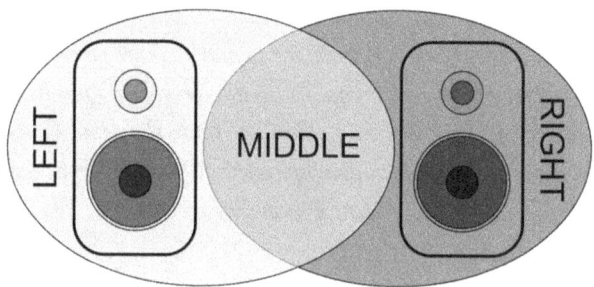

Here's an example of stereo separation. All the sound is placed between the Left & Right channels, like your monitors or headphones.

Mid/Side Separation

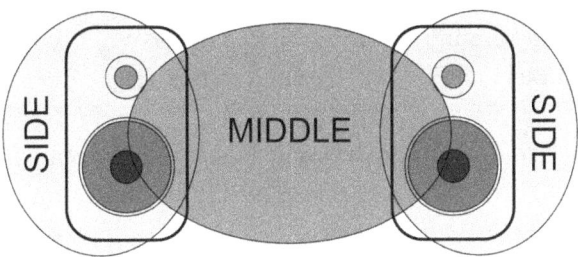

Here's an example of Mid/Side separation. All the sound is placed between the Middle and Side channels. It's a general rule of thumb to place your lead vocal, bass, kick and snare in the center of you mix, with everything else panned out.

Mid/Side Processing

Mid/Side is very helpful during processing. Say you have a lead with background vocals and want your lead to stand out while your backgrounds support the lead with their own flavor and color. We can use some M/S compression to pull this off.

For this example, we're in a mix with 1 lead and 4 background vocals. Here's our routing setup.

- Our lead is routed to Bus 1.
- Our background vocals are routed to Bus 2.
- Both Bus 1 and Bus 2 are routed to Bus 3.

All our vocals are basically being routed to Bus 3. This is where we'll load our M/S compressor.

NOTE: *'Busses' are like literal school busses. The signals in the mix are the students. The driver is the input assignment. The destination is the output assignment. That's the route.*

Bussing
Let's talk about bussing for a little bit.

We can use this routing technique for a few reasons in our mix. First, say some elements may need to be treated in the same way. Let's use our 4 backgrounds vocals from the last example. Working in a DAW, we can put an EQ on all four tracks or we can route these four tracks to an 'auxiliary channel' known as a bus (or *buss*) and put the EQ on that.

Second, using a bus to apply EQ in this scenario would save us 3 devices. That's especially helpful when you're working with a computer and CPU (where the more processors you add, the more load you place on the CPU).

Third, it's much easier to change the settings of one processor on a group of tracks than it is to change the settings of each processor on each track.

On top of those, a fourth reason, is if you had to buy all the hardware versions of your plug-ins. The analog processors would probably cost more than the entire digital studio, all components combined, for good reason. The plug-in will never sound as good as the original, analog unit. This understanding helps explain how the hardware studio world translates into the digital studio world.

In our example, we need a bus for the background vocals so we can treat them as one unit. We in turn route the lead vocal to a bus of its own. We need to treat these two vocal sections differently. Afterward, there may be some adjustments we want to apply to all vocals. That's where the third bus comes in. There are plenty ways we can harness the power of bussing.

Using M/S During Processing

Mid/Side is really a game-changing tool when mixing.
For example, say we have the same mix of 1 lead and 4 background vocals.

With our compressor set to Mid/Side mode, we can compress the vocal in the middle (the lead) and the vocals on the side (the backgrounds) separately while giving them a more cohesive blend as one unit.

For the Mid, we have a ratio of 6:1 with a med-fast attack and med-slow release. We pulled the threshold down to get about -3 dB of gain reduction.

On our Side settings we have a ratio of 4:1, with a med-long attack and a med-long release, pulling the threshold down to get about -5 dB of gain reduction.

With this difference in compression timing between our lead and background vocals, we can create new textures in the mix, giving us a new balance and feel, while keeping the same togetherness (glue).

The more we apply the understanding of the sound field to our projects, the more actualistic and creative the sound is.

❈ 8 ❈
LIMITING

"You matter...until you multiply yourself times the speed of sound squared, then... You Energy." - **Erykah Badu**

THE METAPHYSICS OF AUDIO PRODUCTION

Limiters are another dynamic processor, essentially they are hard-knee compressors with ratios of 10:1 (or higher). After the compression is applied, there's a boost in the output level, then a 'ceiling' is set (for example, -1 dB) to prevent audio levels from overshooting the ceiling point.

In limiters, there's a very short amount of time between the detection of sound and the application of compression (in this case, 'limiting'). This comes in a few different forms.

Traditional Limiting
With traditional limiters, the signal is kept at or below the ceiling level, but sometimes sound can still make it through the ceiling.

Brickwall Limiting
With 'brickwall' limiters, the ratio of compression is set at an extremely high ∞:1 (*Infinity:1*). With such a high ratio, anything that rises even 0.1 dB above the threshold level gets smashed by compression immediately. Really useful with louder material. In harsher circumstances, certain mixes can still peak outside the ceiling.

Soft Clippers
Soft clippers are essentially soft limiters. The term 'soft' comes from the use of a little soft-knee compression at the input stage of the limiter. When audio enters the limiter, the soft-knee compression applied literally softens the attack of the material, making it hit smoother coming out of the monitors. This is done to avoid clipping levels, hence the name, soft clipper.

True Peak Limiting
Most limiters allow some peaking over the set ceiling level. When we take our mix from the DAW to the car or bluetooth speaker, we can hear these peaks create unwanted distortion. True Peak limiters were designed to solve this issue. They limit at the inter-sample level, meaning they limit the samples of the data themselves, as well as the audio coming into the processor. This prevents any analog peaking.

Some Common Limiter Parameters

Now that we know more about the way limiters vary in style, let's talk about some of the usual controls.

Input Gain - controls how much of our signal we're driving into the limiter

> **NOTE:** *With the more gain we add, the more we hear excitement (or possible harshness).*

Threshold - the minimum dB level required to activate the limiter

Ceiling - the highest possible level of signal that can pass through the limiter

Analog & Digital Conversion

One useful concept to understand when limiting is A/D and D/A conversion (pronounced "A to D" and "D to A").

A/D

When we take acoustical energy and record it through a device meant to convert that energy (sound) into data, it's called Analog-to-Digital Conversion (*known as A/D for short*). For example, when we record a vocal using an audio interface, we're using the converters in the interface to convert our acoustical energy into bits of data that our DAW can read and write. This is a crucially necessary step to create digitally.

D/A

Digital-to-Analog Conversion (*known as D/A for short*) can happen a few ways. We'll cover two.

- The first way is when we press play in our DAW. The program plays the bits of data, which is decoded by the audio drivers our recording software. The decoded data/audio is sent through the data cable connected to our audio interface. If you're using USB hubs or any hub, the sound/data is processed through there as well. Once the signal makes it to the audio interface, the D/A converters translate the data into analog audio (acoustical energy).

- The second way is when we take our bounced (encoded) audio files and play them in our car or on a bluetooth speaker. At this point, the data inside the audio file is being decoded by the playback system (CD player, iPod, boombox, etc.) and translated into acoustical energy.

NOTE: *Acoustical energy is a group of audible vibrations produced by a person's voice or an instrument.*

D/A Conversion and Limiting

Remember those pesky 'Inter-sample Peaks'? They cause distortions during analog playback. D/A conversion is when they come into play.

Basically, when you're recording, the converter in your audio interface takes really fast pictures of your sound (so fast that you can't hear one picture change to the next). Each of these "pictures" is called a 'sample'. All the clips and audio files in your session are made up of multiple samples.

For example, say you have a mix that's hitting 0 dB on the master meter and you're not using a true peak limiter. Say the mix sounds good and all you want to do is bounce it. When you take this mix, bounce it (converting it to WAV or MP3) and then play it in the car, you notice that some of the samples inside your audio file have become distorted (as they've peaked above 0 dB). That one really dynamic part you like now sounds too boomy or sibilant. "What happened? It sounded so good in the studio." you say. Meet the inter-sample peak.

True peak limiters compensate for this change in samples and perceivable volume after D/A conversion. With understanding the samples that make up our audio, we can monitor A/D-D/A conversion and use true peak limiting to help our mixes translate better across various playback systems.

❦ 9 ❦
DITHERING

"Turn your side and I'll lift up mine. It'll fit. We good from here." - **That One Uncle In Your Family Helping You Move a Couch**

When limiting audio and dealing with D/A conversion, it's helpful to use 'Dithering'. Dithering is the process of adding noise to offset the displacement of audio (distortion) that happens during the conversion. When we send files via the Internet, it's easier for us to use smaller files. It takes less time to upload and download, hence, the creation of the MP3 and compressed file formats. This type of compression is more related to the amount of storage space an audio file takes up rather than its dynamic range. We can put thousands of songs on our phones because this compressed file format allows us to enjoy the same (shape of) information without taking as much time to access it or using as much space to store it.

There's a term in aviation used when an aircraft gets too heavy and you need to throw something out to lighten it. It's the same effect. There's less cargo in the aircraft (or information in the digital audio file). To save space we sacrifice data. That data equals sound. There's more technical explanation to this, but the short of it is converting from WAV or AIFF to MP3 or AAC will result in some loss of quality, period.

With the knowledge of the way compressed formats work, we can use dithering to offset the loss of data and the artifacts that can occur from D/A conversion. Along with dithering, we use 'Noise Shaping' as an equalizer to offset the random, clashing frequencies from the added noise.

10

SIGNAL PATH (FLOW)

"Empty your mind, be formless.
Shapeless, like water.
If you put water into a cup, it becomes the cup.
You put water into a bottle and it becomes the bottle.
You put it in a teapot, it becomes the teapot.
Now, water can flow or it can crash.
Be water my friend." - **Bruce Lee**

Signal flow is the path audio takes from its point of creation, through whatever processing equipment we're using, to coming out of our playback system.

We're working with a vocalist in this example.

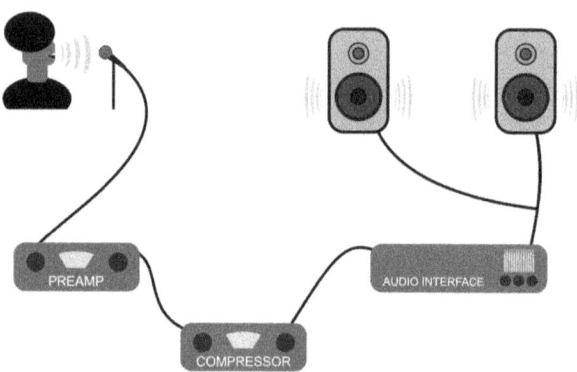

We have a setup with 4 different stages or points of shift in our flow. These devices and the order they are in is what we call our 'Chain'.

For example, in this setup, our chain goes as follows:
Vocalist/Microphone > Microphone Preamp > Compressor > Audio Interface

STAGE 1:
Acoustical Energy comes from the vocalist and goes into the microphone.

STAGE 2:
The microphone converts the acoustical energy into voltage and transmits it as an audio signal through an audio cable to the input of the preamplifier.

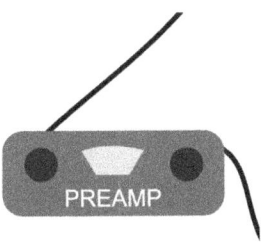

STAGE 3:
The preamp's job is to boost the mic level signal to line level (which is about one thousand times more powerful). If you'e working with a tube preamp, it'll add 'warmth' and 'color' as well.

Warmth is a LMF to HMF roundness that gives feelings similar to songs recorded on vinyl (or the entire 70's). It's that grittiness that makes you feel some type of way.

Color is as the name suggests but more technically, certain sounds and textures can remind you of a specific color or be recognized by your mind as related to that color.

From the preamp, the signal is transmitted out and into the compressor.

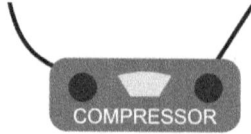

STAGE 4:
The compressor attenuates the signal and then sends it to the audio interface.

STAGE 5:
After these four stages, we hear sound come out the monitors.

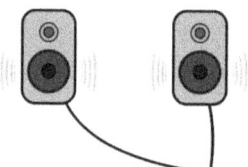

NOTE: *The placement of your microphone in your recording area, the placement of your monitors and their tuning, along with the treatment of the room all have an effect on how your mixes come out. Your best tool is to learn them, meaning get familiar with how your mix sounds in your studio and note the differences when you play it outside your studio. Get familiar with how your favorite songs and songs praised for having a great mix sound in your studio.*

11
GAIN STAGING

"It gets better with time, like fine wine." - **Ron Neal**

Each stage in the chain is important. We need the right amount of energy going into and coming out of each device, to make sure we're getting the best sound possible.

1. The vocalist needs to deliver their performance into the microphone in a way that we can understand all their words and feel their emotions.

2. The signal converted by the microphone needs the right amount of boost for the compressor to have its maximum operating range. This is where the preamp does its job.

3. With enough gain coming in, we can keep majority of the body of the material after compression is applied.

4. The audio interface has a set amount of dynamic range in its input jacks. We want to get the highest signal possible without distortion or clipping, while leaving 'Headroom' for further processing.

NOTE: *Headroom is how much room you have in dB to boost a signal before reaching 0 dB, peaking or distortion. This is why hardware compression is really useful when recording.*

12
DE-ESSERS

"Say it, don't spray it." - **Anonymous**

De-Essers are essentially compressors employing a filtered 'sidechain'.

NOTE: *A sidechain in many situations is the term for a duplicate signal, split from the original, usually for the purpose of alteration.*

With this design, the compression only happens to the filtered frequencies.

In the case of de-essers, the sidechain is operated using filters. Depending on the de-esser, you may be able to choose either a low pass, band pass (*bell*) or high pass filter.

The human voice and the microphones built to record it each have their own frequency responses. Differences in these responses along with recording conditions can cause a common audible distortion in the HF range known as 'sibilance'.

NOTE: *Sibilance is usually not good. It's similar to scratching a chalkboard in frequency.*

Originally, the distorted parts in vocal takes were lowered in volume to blend them into the mix. Later, de-essers were designed as a frequency-based, automated fix for this issue. These devices usually come equipped with Frequency and Threshold controls. Depending on the design, they may even offer compression modes allowing you to change the style of attenuation. When available, two common modes you might find are Split and Wide.

In Split Mode, the filtered signal is compressed when is rises above the threshold.

In Wide Mode, the entire signal is compressed when the filtered signal rises above the threshold.

13

COMPRESSION: DOWNWARD VS. UPWARD

"Step in my square." - **Dallas, TX**

So far as we've discussed compression in this book, we've only approached one form. Though we've covered four different models of hardware compressors all built to work differently, they all lower the signal that rises above the threshold. This is technically referred to as 'Downward Compression'.

'Upward Compression' is the exact opposite. It raises the level of signal that falls below the threshold, while not affecting the signal above it.

It's important to remember that whether we raise the lowest levels of signal, or lower the highest levels of signal, we're still making the range between the lowest and highest levels of signal smaller. This is reducing the dynamic range.

With upward compression, we can bring out the subtleties in a recording without taking away from any of the dynamic parts.

14
GATES

"You call me on the phone, sayin'
Johnny, what's goin' on? Listen
I'll jump off in my ride, quick
'Cause I want you by my side
Now I can hardly wait
'Til I get to your gate
Nothing could break our date
'Cause baby, you take the cake" - **Johnny Guitar Watson**

To round off our discussion on Audio Dynamics in this book, we'll close with the 'Gate'. This dynamics processor is effectively named. It detects a signal and then opens or closes the gate. When the gate is open, there's audio. When the gate is closed, there's no audio.

This can be useful in noisy recording environments, although the best solution for that issue is to acoustically treat and learn your room.

Here are some of the most common parameters found on gates. You might see some you remember.

Threshold - the minimum dB level required for the gate to activate

Range - the maximum amount of dB the gate can reduce the material in volume

Attack - how long it takes for the gate to open

Hold - how long the gate holds the material before starting to release it

Release - how long it takes the gate to release the material

NOTE: *Attack, Hold and Release are all measured in milliseconds (ms) to seconds (s).*

15
EFFECTS PROCESSORS

The Wall of Sound - **Phil Spector, Gold Star Studios**

Now that we've talked about EQ and Dynamics, let's get into some Effects. We'll cover four in this book.

Delay
When you take a signal, make a copy of it and play the copy some milliseconds behind the original, it's called a 'delay'. As the technology developed over time, there have been many different models of hardware delay units, giving us a variety of textures and functions to choose from when working.

Echo
A lot like delay, we hear a repetition of our original signal when we apply the 'echo'. The difference between delay and echo is that with echo, you're taking the original signal and repeating only a section of it. Each repetition you hear is an individual copy of the original signal, played later in time. Instead of playing an entire copy of a signal, later in time, the echo is a repetition of each moment happening in the audio, as it's happening.

Reverb
Though all the tools we use are equally important to the mix, reverberation ('reverb' for short) is one of the most useful when it comes to changing the perceived shape of space a sound is in. Reverb creates width and depth.

Saturation
When we need to liven a signal up because it isn't present enough, we can saturate it instead of adding more EQ or turning the compressor makeup gain up. There are so many saturators available in both hardware and plug-ins. There's usually a tube or circuit made to enhance harmonics involved and a 'Drive' control. Drive determines the amount of signal we're pushing into the tube or enhancement circuit. A virtual preamp would be considered a type of saturator, especially depending on how you use it.

16

MICROPHONE TYPES

"It's cheaper to keep her." - **Johnnie Taylor**

SPL

Before we get into this chapter, let's discuss 'SPL' (Sound Pressure Level). SPL is how much sound pressure a microphone can receive before being damaged. This is a very important system of measurement as we'll be working with many different sound sources and varying levels of sound pressure. For example, we need to be more careful when recording really dynamic vocals and instruments like drums, amped guitars and horns.

Now let's get into a few microphone types to give you a better idea of what you can use in your studio and recording sessions.

Condenser

- A popular microphone in studios, capable of recording a variety of sources, from vocals to guitar to drums
- Very common in recording studios

A highly renowned condenser mic is the Sony C800-G.

Dynamic

- Easy to find in any church or live environment
- Best for dynamic instruments like drums and percussion
- Does not need Phantom Power
- Most durable microphone type

The Shure SM58 is one of the highest selling dynamic microphones.

Ribbon

- Very delicate, older designs
- A beautiful microphone type for vocals that is often heard on music from the 60's and previous recordings
- Easy to damage with high SPL sources

A good example is the Shure 55SH, the one you see the announcer use before a boxing match.

Tube

- Also delicate but more durable than the Ribbon design
- Has a vacuum tube inside that takes time to warm (around 20 mins or so) before giving its best sound quality
- There's a lot of warmth and also room noise with these types of microphones, so be careful not to use too much gain on your preamp

One efficient and highly-rated tube mic is the Manley Reference Microphone.

Choosing Your Microphone

Knowing the types of microphones we can use will help us choose which one is best for each situation. Pick a mic that works well with your vocalist and song.

Microphone Polar Patterns

Equally important as your microphone-to-vocalist choice is your chosen polar pattern. Polar Patterns are your microphone's recording areas. Majority of the mics on the market have one polar pattern. There are plenty brands with different mic designs for different applications. Knowing your mic and its polar pattern will increase the quality of your recordings. Here's a few of the main patterns you can use and familiarize yourself with.

Cardioid - Sound is only captured directly in front of the microphone.

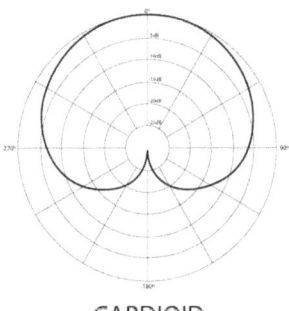

CARDIOID

Figure-8 (*also written as Figure-Of-Eight*) - Sound is captured at the front and back sides of the microphone.

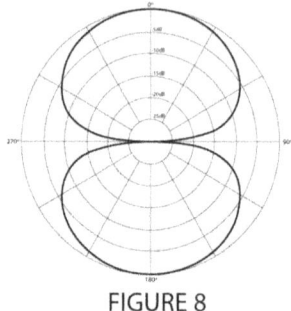

FIGURE 8

Omnidirectional - Sound is captured on all sides of the microphone.

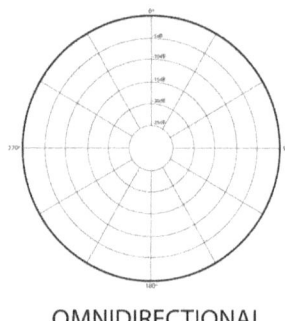

OMNIDIRECTIONAL

There are even mics out there with 'switchable' polar patterns, giving you the capability to choose which blend works best for you and your work.

17
CREATING YOUR CHAIN

"I get around." - **Tupac Shakur**

With all that we've talked about, the time has come to build up our chain. As we discussed before, our chain is a group of signal processors, in a specific order, for the purpose of achieving a desired tonal character and frequency.

Everyone has a unique sound that they're looking for. To achieve that sound, you need a chain tailored to the song you're working on. The devices used in your chain will vary with each piece, changing to suit your needs over time. Check this out.

Microphone > Preamp > Compressor > Interface

In this example, one could say our chain is a preamp and compressor, but for the sake of full understanding, let's include the microphone and audio interface. Say we're using a Neumann U87 as our microphone with an Avalon 737sp mic pre. We follow that up with a Urei 1176 compressor and finish with an Apogee Symphony interface.

THE METAPHYSICS OF AUDIO PRODUCTION

All factors considered, our chain is:
U87 > Avalon > 1176 > Apogee Interface

We have the **Neumann U87** microphone boosted by an **Avalon 737sp** preamplifier. We followed with some smooth compression using our **Urei 1176** compressor. Finally, the enhanced and balanced signal (that's been gain staged properly) reaches our audio interface's input for recording. Build a chain that's vibrant, fun, creative and functional in the long run. It's all about workflow.

18

SOME TIPS

VU Meters are very useful as they monitor signal level over a longer period of time, similar to RMS meters. Very useful.

When working on a track, it's of major benefit to know the sound a processor will give before you add it to your chain. Study your gear. This will save you so much time. Always do a test take. Study how each plug-in or device in your chain responds and hear/feel what the best situation for that response is.

An easy way to stay on top of a recording session is to do some test takes at the beginning. This gives you and the vocalist time to tune your energy. While the vocalist is getting into their song, you can be tweaking your chain. This means setting the parameters on each device so that is flatters the artist's voice and the song. Once you get there, take it from the top, again!

THE METAPHYSICS OF AUDIO PRODUCTION

-10dB v. +4dB
There's more math involved in this, but in short, +4dB (professional grade) is almost 12dB higher than -10dB (consumer grade).

Make sure to gain stage your recording signals or clips before you touch your faders. This will drastically change the dynamic range of your mix. If your DAW has Clip Gain like Ableton's Live, Reason Studios' Reason or Avid's Pro Tools then use that. Apple's Logic Pro X has a similar feature using Flex Pitch. You can also use a free meter plug-in like TBProAudio's VUMeter 2 to get the job done. After you get your audio clips right move to your pan pots, then your faders and then your processors. If you need a specific effect artistically then go ahead and add that. Once you feel good about the gain staging and frequential character of your signals the mix, go crazy with all the effects you desire.

When working with an artist that wants to record while using Auto-Tune, it is best to use Auto-Tune's Low Latency mode. This will lessen the amount of time it takes for your recorded vocal to be processed by Auto-Tune then played back. Have you ever tried this and the vocalist says their voice is coming back in the headphones just a little later than they are speaking? That is latency. Low Latency mode, along with turning down the buffer size in the Audio Preferences, are what keeps the delay in audio at at bay and to a minimum. This will make a large difference in the feel and overall productivity of the session.

Also, when you're using Auto-Tune, make sure it's set to the correct Key and Scale of the song (or vocal performance, if that sometimes differs from the song). After you add Auto-Tune to a track, you can put the Auto-Key plug-in on the instrumental and hit play. As the music plays through the Auto-Key plug-in, it listens and detects the key. Next, just press the Send to Auto-Tune button. This function will change the key and scale of all the instances of Auto-Tune in your session. There's also a mobile app! You need an Antares account to use it freely.

When sending your files for distribution, make sure to use high quality file formats, like WAV or AIFF. Ideally, you want to have your session's Sample Rate set to at least 48,000 Hz *(also written as 48 kHz)*. There's also higher sample rates like 88.2 kHz, 96 kHz and more, but the higher you go the more processing power you'll need in your computer to later mix and add effects. Know and plan for what you're doing. Usually, you'll be sending something like a 24-bit WAV with a 48 kHz sample rate. Converting the sample rate within a session is possible, though not recommended. It's sort of like enlarging a small photo and creating pixelation in the image. The best way to convert the sample rate of your mix, is to record it from one interface set to the original sample rate, into another set to a higher sample rate.

If your microphone has it's own power supply (that you have to plug into the wall then plug your mic into), there is no need for Phantom Power *(48V)* on your preamp.

Eject your hard drives before unplugging them to ensure the safety of your data

Record everything you can, always, even for practice runs. You never know what may happen.

Get creative with your equipment, though leave headroom for further processing. Get the highest, clearest signal possible without clipping and at the same time, leave headroom for processing and effects.

When revising a mix, always Save As. You want to be able to go back and polish specific moments in time.

Plug everything up before you turn anything on.

The bass is the foundation of the mix. If it is off, then there's a high chance nothing else will be on, so make sure to lay the bass properly in the mix so that it supports everything above it.

꧁꧂

Take a picture of the mix in your mind. What does it need? Go from there. Trust your ears, heart and intuition.

꧁꧂

Go with what feels good to your soul and your waters.

꧁꧂

Have a healthy reference for what size and width your mix can be, then go from there with your own creative force and ear for frequency.

Check your mix on more than one playback system. See what it sounds like in the car, on your cell phone, on your friend's bluetooth speaker, everywhere you can, before mastering and release. Understand the relationships between each piece of the music. Ensure each piece has space, is felt and complements the one beside it.

Try checking your lead vocal balance on a pair of headphones, at low volume, ensuring balance between the vocals and instrumentation. This can be very beneficial.

There is no who's got the most hardware and plug-ins competition. Use what you need and what works well for you.

There is what's popularly done, and what sounds/feels good, but there is no "Right Way".
Find your way and what works for you, while having so much fun along the way.

19
OUTRO

It's up to you to find what works best for you and your workflow. Learn the rules, break the rules, reach new heights and set new rules. Use this information to your advantage. Use any perceived or false disadvantage to your real and actual advantage. It's your turn to live and create wonderful works. Make sure to keep your ears open. There are a lot of "mistakes" that can happen, ultimately leading you to greater things within yourself, your world and your work. Do what you wanna do and follow your intuition. You know what's going on. You're the only one with your sonic vision. We're counting on you.

It's been a journey for me to write this and I'm glad it can be a part of your life now. The more you sow into your craft, the more your craft and you will reap. At the same time, it's okay to take time away from the wonder and focus on your personal growth as a being, ensuring you add the best of all your energy to each piece of your work. This self-development and analysis will add so much sauce to your pot, enhancing your entire connection to your creativity when you resume the dinner of delighting in your purpose. Keeping pushing the boundaries of what's "possible". You define what's possible. What do you define? Enjoy each step of the way. Maximize your potential with each breath and moment. Relax into focusing on only that which brings you peace. Put everything into it.

I had so much fun putting this together for you. We have so much more to learn from one another and I'm super excited about it.

Thank you for reading The Metaphysics of Audio Production: A Beginner's Guide to Audio Engineering Foundations, a.k.a. "The MAP".

Much Love and Peace to you all.

- JaRon

SPACE FOR NOTES

ABOUT THE AUTHOR

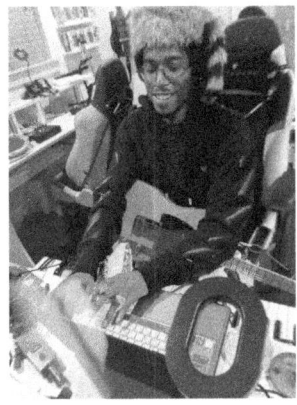

JaRon "The Secret" Adkison a.k.a. "Joe Cool" is a writer, artist, composer, audio engineer, musician and DJ from Dallas, TX, USA.

facebook.com/jaronthesecret
instagram.com/jaronthesecret

www.ingramcontent.com/pod-product-compliance
Lightning Source LLC
Chambersburg PA
CBHW072231170526
45158CB00002BA/840